Jabez P. Dake

Asiatic cholera

Jabez P. Dake

Asiatic cholera

ISBN/EAN: 9783744785532

Printed in Europe, USA, Canada, Australia, Japan

Cover: Foto ©berggeist007 / pixelio.de

More available books at **www.hansebooks.com**

BY

J. P. DAKE, M.D.

Reprinted from Arndt's System of Medicine.

PHILADELPHIA:

BOERICKE & TAFEL.

PUBLISHERS' PREFACE.

THIS book is a reprint of Dr. J. P. Dake's essay on the treatment of cholera in Arndt's System of Medicine. It is undoubtedly the best and fullest presentation of the subject extant. Dr. Dake has been in the thick of every cholera epidemic that has visited this country since 1849, and knows whereof he writes, not only from theory, but from that best of schools— experience.

THE PUBLISHERS.

ASIATIC CHOLERA.

Cholera Asiatique; Asiatische Cholera.

THE term *Asiatic Cholera* is everywhere sufficiently well understood without a display of many synonyms. The disease so designated may, or may not, become *epidemic;* and it may, or may not, display *spasm* or *asphyxia.*

It is indigenous and endemic in India, seeming especially to rise up and prevail in Bombay and Madras, and to reach the commercial tide of the world in the cities of Dacca and Calcutta. It is infectious and, so, indirectly contagious.

History.—It is mentioned as prevailing in India centuries before the Christian era. It was well described, in its epidemic form, by some of the most ancient Sanskrit writers. Characterized by them, according to its leading symptoms, it was known as *Vishû dschikâ*—vomiting and sweating; again as *Alsikâ*—cramps; and again as *Rilambikâ*—collapse. Its first appearance in Europe was in the year 1830, coming through Russia and Poland. Observing no definite period, it has repeatedly ravaged the countries of Europe, scourged the British Islands, and crossed to America since that time. The first fatal case in Russia was at Orenburg, August 26, 1829, from which time it passed from town to town, all through the year 1830, not reaching St. Petersburg till June, 1831.

It was brought to England by merchant ships from Riga, quite a number of fatal cases occurring on the river Medway during the summer and autumn of 1831. Its first appearance in epidemic form was at Sunderland in November, whence it spread to Newcastle, Edinburgh, Glasgow, Belfast, etc., arriving in London, February, 1832.

First Visit to America.—From Dublin it went to Cork, and thence to Quebec in America, where it appeared in June, 1832. It is a question, not fully settled, how the disease came to New York city, whether overland from the St. Lawrence, or by ship from infected ports abroad. It was recognized in that city before the end of June; and by the first of September it was spreading along the Mississippi. Before the end of October, it was in New Orleans.

It prevailed in Kentucky in 1833, and reappeared on the St. Lawrence in 1834, not spreading extensively however. And again, in 1835, it seemed to have been imported from Cuba into New Orleans and Charleston. It has been claimed by some writers that there was no fresh importation during the years 1834 and 1835, but rather, a springing up from a few temporary depositories of the seed.

Second Visit to America.—After the lapse of nearly eighteen years, in July, 1847, the cholera again entered Russia, at Astrakhan, and, ascending the Volga, pursued nearly the same line of March taken in 1829-30, reaching Orenburg in October. In the same month, it appeared in Constantinople, whence it passed to Odessa and up the Danube into Austria, and thence into Italy the following year.

During the winter of 1847 it was on the Baltic and also in the interior of Russia, not severe, but quite sufficient to demonstrate its presence.

It was at Berlin in July, and London in October, 1848. So far as known, the first vessels from infected ports, having had cholera cases aboard, arrived in American waters in the winter of 1848, the one landing at New York, December 1st, and the other at New Orleans, December 11th. At the New York Quarantine about one hundred cases occurred, one-half being fatal, while in the city proper there were only two cases. There was no spread of the disease, so far as recognized, till the following April, during which time other vessels had arrived with emigrants from infected ports. The first case in the city occurred on the 11th of May, from which time the disease prevailed, with varying severity, till the close of September, causing over five thousand deaths. The vessel that had brought cholera to New Orleans, December 11th, had lost thirteen of her passengers on the way. One of her emigrant passengers died with the disease in the city two days after her arrival, no quarantine having been enforced. Before the end of that month four hundred deaths from cholera were reported, and in January six hundred. The number of fatal cases continued to increase till June, when it reached two thousand and five hundred. A steamboat, having had several cases among passengers and crew, arrived at Memphis from New Orleans, December 20th. Two days afterward, the first case appeared, near the wharf; and four days thereafter two more, on a flat-boat near by, all fatal.

After the lapse of nearly three weeks, during which time other deaths occurred among persons frequenting or residing near the landing, the disease began to appear in the interior of the city. Another steamboat from New Orleans, having had nine deaths on board, from cholera, arrived at Nashville, on the Cumberland river, December 27th ; but no cases occurred among the residents of the city till the 20th of the following month, after the arrival of several other boats from New Orleans. Soon after that date the disease became epidemic. Thus, from New York and New Orleans, both, the cholera spread till it had again traversed the busy haunts of the United States, as an epidemic. The river and lake cities suffered greatly, being on the lines of greatest and most speedy travel.

It was the fortune of the writer to first encounter the disease at Pittsburgh, at the head-waters of the Ohio, where its ravages were severe during the months of June, July, and August, 1849. The constant arrival of emigrants from infected ports, during the winter of 1849–50 and, possibly, the survival of the seed from the previous season in the warm and unclean parts of New Orleans, caused the disease to appear again in that city and in the river towns above, during the summer of 1850. It extended north and eastward only so far as Pittsburgh.

Third Visit to America.—During the years 1851-2-3

cases of cholera occurred and the disease was epidemic in some parts of Europe, but broke out with greater fury in Russia, in 1853, extending westward into Germany, southward into Austria and Italy, and northward along the shores of the Baltic, to Denmark and Sweden. From July to December there were twelve hundred cases in London. It was severe in the leading cities of England and Scotland, and lingered about them, more or less, during the winter of 1853–54.

Some twenty-eight vessels sailed from infected ports to the United States before the following summer, with a total loss of over eleven hundred passengers with cholera while on the way ; and yet, the first cases of cholera reported in this country in 1854 occurred in Chicago, where emigrants unpacked and brought into use the clothing and bedding from infected places abroad.

Though cases of cholera, usually termed sporadic, had occurred in different parts of the country during the years 1851–2–3, it is generally believed that there was a fresh importation of the seeds in 1854. Appearing in Chicago in April, it was not severe till June, when it was declared epidemic. In May it was in Detroit, and by June had reached New York. It had appeared, also, during the winter of 1853–54, along the Mississippi, becoming especially severe at St. Louis, from February, all through the spring and summer. Thence it spread in all directions, especially along the rivers traversed by steamboats. It was severe at Pittsburgh from the middle of September to the 10th of October. It was in Nashville in June.

As before, the disease was confined to the chief lines of travel, appearing to go by stage-coach and wagon, as well as by steamboat and railway train. It gradually disappeared as the cold weather came in the winter of 1854–55.

Fourth Visit to America.—During the year 1865 the cholera was very severe in the Bombay presidency, eighty-four thousand people dying with it. Thence it extended along the Red sea and came to Alexandria with the pilgrims from Mecca. Conveyed by the shipping of the Mediterranean, it reached Marseilles in June, and passed on to Paris, where it raged, at times with great fury till the end of the year. In October and November ships, having had cholera cases aboard, arrived in New York, from Havre. Cases occurred on Ward's Island, but the disease gained no headway during the winter months. In April, 1866, a steamship from Liverpool arrived at New York, having lost forty-six passengers with cholera before reaching Halifax, where she was held in quarantine and subjected to fumigation. Another steamship from Liverpool arrived, April 18th, having lost thirty-eight persons with cholera; and, during the spring and summer, at least half a dozen more vessels arrived there, reporting deaths from cholera on the way.

The first cases in New York occurred early in May, and in quarters usually frequented by emigrants and seamen. Great precautions were taken by the physicians and health authorities, so that only individual cases appeared.

In the month of July, in a quarter of Brooklyn inhabited almost entirely by foreigners, a severe epidemic occurred; and a little later it was severe, also, on Ward's Island. Extending to Governor's Island, and other points, it was soon distributed over the country. Before the end of July it was down the Atlantic coast and along the Gulf to New Orleans; and before September it was in Texas, and up the river in Mississippi, Kentucky and Ohio. Early in September it raged severely at Nashville; and by the middle of November had apparently disappeared from the country. A notable feature of the visitation of 1866 was that it seemed to cling very closely to army quarters, and to proceed with troops in their travels from place to place. But that feature was not peculiar to the epidemic of 1866, as in all epidemics in the old world the cholera had shown a great partiality for armies, caravans, crowds of pilgrims and for extemporized settlements.

In the summer of 1867 cholera reappeared among the troops in the west, but did not prevail to any extent among the residents of the country.

Ships from infected ports came to New York, having lost passengers with it on the way, but the careful measures of quarantine, or other favoring circumstances, prevented an epidemic.

Fifth Visit to America.—From its eastern haunts cholera invaded Russia again in 1869, and spread gradually over the empire. Thence it went into Turkey, Austria and Italy southward, and westward into Germany, reaching Havre and London in 1872. It gained no foothold, however, in England, but passed on westward to Brazil and the United States. The strict quarantine at the port of New York allowed no spread of the disease there; but it is believed that the seeds were carried through from the ships in the unopened baggage of emigrants from infected ports, as the disease appeared at several points inland to which they had gone. The initial cases were in Louisiana, February 9th, 1873. In April there were some in Mississippi, Arkansas and Tennessee; and in May in Illinois, Missouri, Kentucky, Indiana and Ohio. The disease was epidemic at Nashville in June. It was not promptly recognized on account of the bilious character of the stools in the early cases. In that regard it was much like the form of cholera that prevailed in Japan a few years later, and with which 158,000 people were said to have perished in one year.

Enough of history has now been given to show something of the nature and ways of cholera, and of the measures requisite for its prevention.

Ætiology.—Much has been written, especially during the last fifty years, upon the essential cause of Asiatic cholera, and various have been the theories put forward in regard to it. It has been looked upon as some mysterious influence, from the conjunction of certain planets, from chemical changes in the composition of the air, from vapors arising from the earth, from pollution of water and from other more or less inscrutable causes. It has been looked for in winds, earthquakes, floods, swamps, cesspools and sewers. The scalpel, the microscope and chemical reagents have been employed in the search. Of late years,

writers on the subject have been almost unanimous in
regarding East India as its birthplace, and the field
producing the seeds that, under favoring conditions,
have given rise to the epidemics which have swept the
greater portions of the habitable globe.

It must be mentioned, as a singular fact in the literature of this subject, that Hahnemann, before he had
ever seen a case of Asiatic cholera, while he was almost
hidden away at Köthen, from his careful reading of the
reports coming to him from the fields being traversed
by the destroyer, put forth views as to the essential
cause of the germs of the disease, which are to-day
indorsed and confirmed by the results of the most
scientific research.

This is his language (*Lesser Writings*, p. 758): " On
board ships—in those confined spaces, filled with
mouldy, watery vapors, the cholera miasm finds a
favorable element for its multiplication, and grows into
an enormously increased brood of those excessively
minute, invisible living creatures, so inimical to human
life, of which the contagious matter of the cholera
most probably consists—on board these ships, I say,
this concentrated, aggravated miasm kills several of
the crew; the others, however, being frequently exposed to it, at length become fortified against it, and no
longer liable to be infected. These individuals, apparently in good health, go ashore, and are received by
the inhabitants without hesitation into their cottages,
and, ere they have time to give an account of those
who have died of the pestilence on board the ship,
those who have approached nearest to them are sud-

denly carried off by the cholera. The cause of this is, undoubtedly, the invisible cloud that hovers closely around the sailors who have remained free from the disease, and which is composed of probably millions of those miasmatic, animated beings, which at first developed on the broad and marshy banks of the tepid Ganges, always searching out in preference the human being to his destruction, and attaching themselves closely to him, when transferred to distant and even colder regions, become habituated to those also, without any diminution either of their unhappy fertility or of their fatal destructiveness."

It would not be easy for Dr. Pettenkofer or Dr. Koch, after all their investigations in this field, with all the improved appliances of modern times, and all the accumulated lights of observation and experience, to set forth more clearly and satisfactorily the specific cause of Asiatic cholera, its home, its haunts and its ways, than was done by Hahnemann more than fifty years ago.

If the views put forth by these modern scientists, now so favorably received throughout the medical world, have led to any modes of recognition, or measures of prevention and cure, better than those suggested by the sage of Köthen, no proofs of the fact have yet appeared.

The infectious character of cholera, and its transportation from place to place and person to person by reason of the " minute, invisible, living creatures," constituting an essential cause, was ridiculed and violently opposed by such writers as Hufeland. Only after the accumulation of facts, developed by many epidemic years of cholera, has it come to be acknowledged generally, among authors as well as practitioners, that the disease does not spring up *de novo* from beds of filth under summer heats, in the several countries where its ravages have been felt, without the access of the living germs, indigenous only to India.

The following ætiological propositions may be considered as quite generally admitted now among well-informed medical men:

I. That the essential cause of malignant or epidemic cholera is indigenous to East India alone.

II. That it consists of minute organisms, living and susceptible of rapid multiplication, as yet identified only by their effects in the human body.

III. That these organisms occur in the ejections of persons already infected, and may be potential in a moist state, in water, food or filth, or in a dry state in clothing, bedding, merchandise and confined air.

IV. That their vitality, especially in a dry state, may be retained for days and, possibly, for weeks.

V. That persons become subject to their influence chiefly through the alimentary canal, and less often by other channels.

VI. That their potency is impaired or destroyed by a very low or very high temperature; also by the processes of fermentation in their menstruum; and, again, by chemical reagents.

VII. That their actual power depends on their number and organic integrity within the human body, and, consequently, their effectiveness upon the receptivity or non-resistance of the individual subject.

This last proposition, touching upon individual pre-
disposition or liability to cholera, calls for an honor-
able mention of the advanced and wise thoughts of
Hahnemann. He said (*Lesser Writings*, p. 759): "The
sailor closely but invisibly environed by the pesti-
ferous, infectious matter, against which, however, as
has been observed, his own individual system is, as it
were, fortified by long resistance of his vital force to
its action, and by being gradually habituated to the
inimical influence surrounding him—this sailor (flying
from the corpses of his companions on board) has
often gone ashore apparently innocuous and well, and
behold! the inhabitants who hospitably entertained
him and, first of all, those who came into immediate
contact with him, quite unused to the miasm, are first
most rapidly and most certainly attacked without any
warning and killed by the cholera, whilst of those
more remote such only as are unnerved by their bad
habits of life are liable to take the infection."

These graphic words advocate the truth that new-
comers, persons in health, fresh and receptive, are the
first to take the disease, and to take it most severely.
Persons responsive to other influences, sensitive to
external impressions generally, and by no means necessarily sick from other causes, are the ones who take
it most readily.

Beside the essential and predisposing causes of
cholera, there is another class to be considered—the
exciting; and, as belonging to this class, the following
may be mentioned:

1. Fear, disturbing the harmony and efficiency of

organs, especially hindering the processes of digestion and assimilation.

2. Improper food and drink, burdening the stomach and exciting undue alvine evacuations.

3. The impression of cold upon the surface of the body and limbs, or within the stomach, disturbing the circulation, and so impairing the strength of vital resistance in general.

Pathology.—Inquiries into the natural history of Asiatic cholera, as manifested in its subjects, in order to learn upon what tissue its first impression is made, and how it proceeds in the disturbance of one organ after another till dissolution results, must always be a matter of interest to the thoughtful physician. While the symptoms or signs of the disorder, in this as in other cases, must mainly serve as the practical guide in therapeutics, a knowledge of the interior changes, the structural lesions, corresponding with the outward display, must afford aid in treatment as well as in diagnosis and prognosis. The undue importance, however, given to pathological views by disappointed therapeutists of the old school, has led more to useless disputations than to any solution of practical difficulties. At the very time that Hahnemann, under the guidance of the homœopathic law, was pointing out the great remedies for cholera, his distinguished neighbors, Andral and Broussais, of Paris, were hotly contending, the one for the neurotic, and the other for the inflammatory, theory of the disease.

Andral, on his theory, argued for antispasmodics, while Broussais, on his, favored antiphlogistics ; and yet no practical good came from either source.

Numerous post-mortem examinations have shown that in cases attended with much vomiting and purging, especially the latter, the epithelium along the alimentary canal, particularly in the small intestines, is destroyed in patches.

The same has been observed, to some extent, in the urinary tracts. Hayem, of Paris, after a series of examinations, sums up his observations in the following words : " The only organs constantly involved are the intestines. The capillaries, the different layers of the epithelium, the sets of glands, and the villi, had all undergone certain changes, but differed in no way from the changes observed in ordinary intestinal catarrh. In the blood was found an increase of the white corpuscles and small fragmented globules. These are explained by the stasis of the blood in the algide period, and the decrease in the proportion of water. No microscopic characters peculiar to cholera were found."

On the other hand, Dr. Koch, in his examinations made in Egypt, India, and later in France, claims to have discovered a species of microbe in the dejections and the small intestines, peculiar to cholera victims. He asserts that this parasite, denominated by him *the comma bacillus*, is present in all cases of Asiatic cholera.

But, so far, he has failed in the proof of its specific causative character by originating a case of cholera through its agency, when entirely dissociated from other probable causes. If, as claimed by Koch, his bacterium cannot exist in a dry state—if its vitality depends on a moist menstruum or nidus—he must account for, what has been repeatedly observed, the conveyance of the seeds of cholera in the baggage and bedding, and confined air of chests, taken inland to distant points from infected ships. If all he asserts regarding the habitat of the comma bacillus be true there is strong ground for the belief that it is not the specific cause of epidemic cholera.

When asked in what way his bacillus acts in the production of cholera, Dr. Koch replied: " First its pressure causes diarrhœa and vomiting, leading to thickening and chilling of the blood ; then, it secretes an intoxicating poison which causes the dry, instantaneous cholera without diarrhœa."

It is not easy to understand how enough bacilli to secrete the " poison " necessary to cause dry cholera could be present without occasioning the " pressure " that he claims to be the cause of vomiting and purging.

This discovery by Dr. Koch has been heralded over the world as one of great importance, while, in fact, it places the pathology of Asiatic cholera, as well as its prevention and cure, not one step in advance of where it was when that disease first invaded Europe and America.

The world may be thankful that the recognition of the destroyer from the Ganges does not depend upon the microscope, nor its successful treatment upon anything that the theories of Dr. Koch may suggest.

Symptomatology.—The symptoms of cholera can best be set forth by mention of typical cases.

A., being in the enjoyment of usual health, suddenly has a disposition to stool. The discharge from the

bowels is painless, or nearly so, quite watery, of light
color like thin, flocculent rice-water.   A second action
occurs, with a sense of relief, as though the evacuations
would be beneficial.   Repeated, gushing and more copious discharges occur, with a feeling of weakness,
nausea, and epigastric distress.

Vomiting and purging are simultaneous ; there is
great thirst, yet a prompt ejection of all liquids drank ;
cramps come in the muscles of the legs, arms, and ab-
domen ; the voice is husky and cavernous ; the surface becomes cold and wet with perspiration, then
shrivelled and blue ; the urine is scanty or suppressed ;
the pulse is weak at the wrist ; the face shrunken and
old-looking ; there is great restlessness and tossing ;
loss of voice and pulse, collapse and death.

B. has been well, except, for a few hours he has had
a sense of fullness in the head, perhaps a ringing in
the ears, when, suddenly, he has nausea, vomits food
eaten hours, perhaps days, before ; copious watery
stools, at first dark, and then like rice-water, occur ;
vomiting and purging are simultaneous ; no pains, no
cramps, but rapid collapse and death.

C. has had some dizziness, and feeling of weakness,
but continues at business till, walking in the street, he
is suddenly seized with cramps in his legs, falls, and is
picked up ; his breathing is difficult, his face livid ; he
sinks rapidly, becomes pulseless, and dies within an
hour, without the least vomiting or purging.

D. retired, after drinking beer, or lemonade, or buttermilk, feeling quite as well as usual, and is awakened
toward morning with nausea and vomiting ; one very

copious dejection occurs; there is great epigastric distress, more vomiting, collapse, and death, all in the
space of two hours.

In different persons these symptoms are varied;
some having no purging, and others no vomiting or
cramps. Some sink and go into collapse quickly,
while others resist longer, and maintain the fight for
twenty-four hours or more. Some have little or no
pain, while others have cramps, not only in the limbs
but likewise in the muscles of the trunk, and even of
the face.

The pathognomonic symptoms of Asiatic cholera
may be stated briefly as these :

1. Rice-water dejections, very copious.
2. Sudden and excessive vomiting and purging.
3. Cramps in the legs, later in the arms and trunk.
4. Husky voice—sepulchral, choleraic voice.
5. Leaden hue of the skin—later, shrivelled and cold.
6. Sense of heat while cold to the touch.
7. Rapidly sunken eyes and pinched features.
8. Suppression of urine.

Persons may die of heart failure through fright, and
from apoplexy and convulsions from other causes,
during a cholera epidemic; but none will die with Asiatic cholera without exhibiting three, or more, of the
eight symptoms just mentioned—in a large majority of
cases all of them. As shown in the typical cases pre-
sented, there can be no classification of symptoms by
stages, each stage coming in a particular order, and
occupying a definite period of time.

Cramps may come before vomiting, and vomiting
before purging, and *vice versa*.

In cases of recovery, the symptoms mentioned become less severe, and less frequent, till convalescence is secured. The vomiting and purging cease, the cramps relax, warmth returns, and the patient is safe. A darker color of stool and reappearance of urine are favorable.

Diagnosis.—The recognition of a case of Asiatic cholera must depend on the symptoms displayed; and yet, without its history, showing the nearness of some case or cases of that disease, it might, at times, be no easy task. Cases of sporadic cholera or cholera morbus occur every summer, presenting symptoms quite similar to those given as characteristic of Asiatic cholera; but usually the vomiting is less forcible, the stools less gushing and of darker color, the voice not so husky, the skin less blue and shrivelled, and all of these less rapid and fatal.

Should the microbe of Koch prove to be an invariable accompaniment of Asiatic cholera, always discoverable in the ejecta of the subject, the differential diagnosis might become quite sure to the expert microscopist, provided always, that the same microbe was not discoverable in the ejecta of persons having cholera morbus or other enteric and gastric affections. At present we must take in the history and relations, as well as symptoms, of first cases in order to arrive at a correct diagnosis.

Prognosis.—That there is a difference in the morbific power of the germs or seeds of Asiatic cholera, as shown in the same localities in different years, and in different localities the same year—that in one locality or during one season, the cases are more rapid and more fatal—there can be no question. The explanation of this fact has been variously based on differences in rainfall, in temperature, or in weight of atmosphere. Aside from such extraneous influences, affecting the people, there is good reason to suppose that there may be differences in the vigor and propagating readiness of the germs themselves, arising from causes with which we are, as yet, not at all acquainted. The prognosis, then, in this disease cannot always be the same ; nor can it be based on anything that we may know beforehand of the individual or his surroundings, except as already mentioned in the section on predisposing and exciting causes. It may be said, however, that the prognosis is more favorable in winter than in summer, and toward the close than at the beginning of an epidemic; also, that it is more favorable in persons not addicted to the use of stimulants and narcotics.

Treatment.—The measures resorted to in the prevention and treatment of cholera have been as numerous and various as could be devised by the learning and ingenuity of man.

Empiricism has gone the rounds and tired itself, and theorists have taxed their skill in vain, to furnish satisfactory remedies. Sad as the results have been where cholera has prevailed, no one can read of the ups and downs of vaunted specifics, of the worthless and often disgusting trash, and of the wild fancies that, one time and another, have had the confidence of the profession as well as of people, without a feeling of amusement.

The leading medical journals of the old school abound in acknowledgments of failure after the coming and going of each epidemic; and so, likewise, do the voluminous reports of governmental commissioners and inspectors.

After the epidemic of 1873, in the United States, a surgeon was detailed by the War Department to inquire and report upon it. After giving its history and noting the various lines of treatment, with comparative results, he says: " In the advanced stages of the disease the entire range of the Pharmacopœia seems to have been brought into use with no better results than have been obtained in previous epidemics." The number of cases reported to the surgeon was 7356, with a mortality of 52 per cent. ; but this rate was moderate, compared with that of the epidemic of 1884 in France and Italy, which ranged from 50 to 90 per cent.

When it is remembered that all the lights of medical science in the hands of a profession, regulated by governmental boards and censors in those countries, failed to save more than one in four or five of the people attacked by cholera, the conclusion is irresistible that the boasted lights were worthless and the censorship a fraud, or there was some great obstacle in the way, some unusual hindrance to medical ministrations. When we read in the daily bulletins of the consternation among the people, how they not only had no confidence in what medical men could do for them, but were so possessed of fear that the doctors would make them worse by their experimental drugging that they actually drove them away by peltings with sticks and

stones, we may begin to realize how little of promise
was afforded by orthodox medicine in the large cities
of those enlightened countries.

Where authority is given to official boards to regu-
late the practice of medicine there is, necessarily, an
orthodoxy, a fashion, sustained by law as well as cus-
tom, that is deaf to the criticism of heterodoxy, and
intolerant of every proposed measure of medical re-
form.

At Naples, where more than six thousand people
died with cholera, during August and September, 1884,
Dr. Rubini, a veteran of more than four-score years,
with a record of success in cholera treatment equalled
by that of no other man, appealed to the authorities in
vain to provide *Camphor* for the people, a remedy
which then, and often before, in his hands and the
hands of others, had proved itself an almost infallible
preventive. Speaking of this, Dr. Cigliano says :
" But it was not to be ; and the welfare of the people
has been sacrificed in obsequious obedience to Sar-
dinian officialism, which is as careless of public grati-
tude now as it was in 1854, when Signor Giustiniani
was forbidden to found, at his own expense, a hospital
for the homœopathic treatment of cholera at Genoa! "

Prevention.—The failure in measures for the cure
of persons attacked with cholera, on the part of tra-
ditional medicine, has given especial fervor to efforts
for the anticipation and prevention of that disease.
The surgeon commissioned to report to the United
States Congress on the cholera epidemic of 1873, said
(p. 15): " With the admitted uncertainty of therapeutic

measures in this disease its outbreak at sea should be
the signal for the most scrutinizing search for its ori-
gin, with a view of thence stamping it out." And, in
concluding his recommendations, he said (p. 19):
" The true remedy against cholera is preventive medicine."

These quotations plainly show the utter poverty of
old-school therapeutics, and the urgent desire for prevention, as realized of late years in the highest quarters of that school.

In the new school, from the time of Hahnemann,
there has been a full understanding of the importance
of preventive measures, such as isolation, disinfection,
hygiene, and prophylaxis.

The first publication upon cholera, from the pen of
the master, issued in pamphlet at Leipsic in the autumn
of 1831, contained strong and conclusive arguments
upon the infective character of the disease and the
necessity of isolation and disinfection. Clinical experience and earnest research have at last taught the
medical world that he was right and Hoffman wrong.

Quarantine.—As cholera has never appeared in
America and become epidemic without having shown
itself, shortly before, in European countries with which
we are in constant and comparatively close communi-
cation, the first step toward its prevention must be
taken by the general government. Thorough inspection of all vessels coming from infected ports, and
satisfactory proofs of no cases of cholera in transit,
should be required before passengers and crew, and
baggage and even merchandise, are allowed to land.

Detention and inspection, by United States authority, should be practiced upon all vessels coming to us during, and for months after, the prevalence of cholera in European cities; and where cases of cholera have occurred upon any of them while at sea, thorough disinfection, as well as detention, should be enforced. All baggage and clothing and bedding and the cargo, so far as exposed to the cholera miasm, should be removed, opened out, and subjected to dry heat and the fumes of sulphur in close apartments, and then sprinkled with a strong alcoholic solution of camphor before being removed from quarantine. And the detention of passengers and crew should be long enough to allow the development of cholera in any who may have taken in the seeds of the disease.

Public Hygiene.—As the foothold and prevalence of cholera in a community depend somewhat upon conditions that man may control, such as the quality of air, of water, and of food, the second step toward prevention must be taken by the local authorities. The water supply should be carefully inspected and improved where subject to pollution; springs, wells, and cisterns, receiving sewage or filthy surface water, should be closed; areas, vaults, cesspools, alleys and streets should be cleaned, and, where necessary, disinfected by special means. While no amount of impure air, or water, or food can produce a case of Asiatic cholera, they can afford a menstruum, or nidus, in which the seeds of that disease may germinate and multiply, and so beget an epidemic.

Private Hygiene.—When quarantine has failed to

keep cholera out of the country, and public hygiene
has done no better for the community, individual action
is imperative.

Those residents who have a particular fear or personal apprehension of cholera, and those who are subject to diarrhœa, should leave upon the sure approach
of the dreaded disease. They, and such others as may
not be needed to attend upon the sick, or to supply the
necessaries of life for the community, should take ref-
uge in localities away from the usual highways of
travel and commerce; and arrangements for such re-
sorts should be made in advance of the first home
cases, in order to avert a panic, and to avoid the distribution of the disease.

A community thus relieved is in better shape for the
encounter, as physicians will have more time to look
well after those who remain and are stricken down, and
there will be less terror and neglect generally.

For those who remain, the following suggestions
may be useful :

1. Make no change in clothing, except to put on
something a little warmer, especially a bandage of
flannel secured tightly around the body, extending
from the arm-pits downwards.

2. Avoid sitting and, yet more, sleeping in a draft
of air ; stay in at night, and take the usual amount of
sleep.

3. Take no unusual baths; especially avoid bathing
in stale or stagnant water.

4. Take no unusual exercise, but attend to accus-
tomed business.

5. Attend no meetings in chilly or damp weather, except in rooms properly warmed; and, in general, avoid groups of people where the subject of cholera is being discussed.

6. Keep away from rooms and houses in which there are, or lately have been, cases of cholera, unless compelled by duty to be there.

7. In diet make no great or sudden changes, except to avoid articles likely to induce diarrhœa, such as cabbage, cucumbers, roasted green-corn, beets, turnips, squashes, new potatoes, snap-beans, etc. Most persons do well to adhere to bread, toast, crackers, rice, beef, breakfast bacon, well stewed tomatoes, and tea, cocoa, or coffee. Avoid articles, usually wholesome, that may be the vehicles of infection, such as milk (except when just boiled), buttermilk, cheese, etc.; and all articles of food or drink while in apartments cccu-pied by cholera patients. Veal, poultry, fresh fish, and eggs are not safe. Ice-water, or large draughts of cold water, calculated to chill the stomach, are bad. The safest water is from cisterns that are allowed to take in no rains after the close of winter. Distilled or fer-mented drinks are unsafe; though it is not best to make too great or sudden changes in their use—espe-cially not to increase them in quantity or frequency.

Fruits, especially uncooked, should generally be avoided. Much good may be done by relief commit-tees in the distribution of rice, crackers, bread, beef, bacon, and tea and coffee, among the poor, who are largely dependent on cheap vegetables for nourishment, especially during a cholera epidemic, when their occu-pations and incomes are suspended.

Prophylaxis.—The homœopathic rule applies not only in the use of remedies for the sick, but likewise in the use of agents for the prevention of sickness in the healthy. This fact has been shown with great clearness during the prevalence of cholera, and is of too great importance ever to be forgotten or misunderstood. Without an effort at argument here, this proposition is submitted, that so long as the institution of a similar pathological condition is necessary for the removal of one already existing, a successful prophylactic must have the power to institute in the healthy organism a condition similar to that to be prevented. Vaccination for small-pox, Belladonna for scarlatina, and Camphor and Cuprum for Asiatic cholera, afford good examples. Whether the prevention comes from the exhaustion or molecular change of a stored material, or a tissue, the presence and original state of which constitute susceptibility in the individual; or whether a special power of resistance is aroused or developed, or a degree of tolerance affected, whereby the germs or morbific influences are made harmless, it is clear that an abnormal state is occasioned, and that that state is similar to the one effected by the disease to be guarded against.

Persons in the most robust health, with each susceptibility normal, or excited by a predisposing cause, such as already mentioned, are the ones first to take the cholera, and to have it most severely. Those, as mentioned by Hahnemann, gradually inured to the miasm, or subjected to the influence of a homœopathic agent, like Camphor or Cuprum, are exempt or slightly affected by the disease. This fact we accept, whether its explanation, so far as made, be satisfactory or not. In the tract issued by Hahnemann in 1831, as the cholera was approaching from Austria, directions were given for the use of Cuprum as a prophylactic agent. In the "*Cajeput oil*," imported from India in copper vessels, he recognized "a Camphor property as well as some portion of copper;" and, knowing that these agents were capable of inducing symptoms closely resembling those of cholera, and, further, that the Cajeput oil had proved very efficient in the treatment of that disease, he advised the use of Cuprum as a preventive, and Camphor as the first and most important remedy.

Copper plates, and chains, and rings had gained favor in Hungary as cholera preventives.

The ready and correct perception of the truth by Hahnemann, in the ætiology of Asiatic cholera, had its parallel in his ready and vigorous grasp of the truth in its prophylaxis. The indorsement of his views as to the former, throughout the medical world, is now nearly equalled by the sanction given to his recommendations as to the latter.

No prophylactic agents for cholera have been so successfully employed, during any epidemic in any country, as *Cuprum* and *Camphor*.

During the epidemic of 1884 in Naples, the latter was in greatest favor. Dr. Cigliano says: " The prophylactic which has enjoyed our entire confidence here (Naples) during the past epidemic, and also in those

which preceded it, is Rubini's Camphor,* given in
drop-doses, two or three times a day. A very large
number of persons have used it, about fifty thousand,
and all, with the rarest exception, affirm that they
have been preserved by it, though they have nursed
cholera patients, and have lived in houses wherein
people have died with the disease."

Camphor was freely distributed to the poor from the
Homœopathic Central Pharmacy of Naples, when the
municipal and all other authorities refused to provide
it for the people.

But, outside of Italy, Cuprum has been more frequently employed than Camphor, as a preventive. In
the epidemics of 1831, 1849, 1854, 1866 and 1873, it
made an excellent record wherever homœopathic
remedies were brought into use.

Nor has all the favorable testimony come from the
avowed advocates of homœopathy. Dr. Burq, of
Paris, who had been investigating the therapeutic properties of metals variously applied, made the discovery
in 1849 that workers in *copper*, foundrymen, machinists
and others, had experienced a wonderful exemption
from cholera. In reporting his observations to the
Academy, in Paris, he said: " The preventive effect
was, no doubt, produced *directly* by contact, and in
proportion to the amount of the protecting metal, and
indirectly by simple vicinity, as in the case of those
who are near a lightning-rod ; at least, it is by the latter
mode only that we can account for the marked pre-

---

* A saturated solution made by trituration of the gum in alcohol.

servation which was experienced by the neighborhood of nearly all the copper foundries, unless it may be attributed to the emanations from the metal, caused by its fusion, or, rather, by its manipulation in the workshop, either in the form of highly-attenuated particles, or of effluvia of a peculiar character."

Dr. Burq was inclined at the outset to the opinion that the electric or magnetic properties of copper and its alloys should be credited with the protecting power; though, as intimated above, he finally considered it possible that the inhalation and absorption of " highly attenuated particles " might have something to do with it.

After saying that the protecting power might reside in other metals, in proportion to their electric and magnetic similarity to copper, he said : " The *curative* power appears, on the other hand, to reside in copper alone, which would seem to act upon the cholera miasm as sulphate of quinine does upon intermittent fever. This curious property has been very often brought to our notice, attested by the most incontrovertible evidence ; and many a workman or master of a copper foundry has been preserved from cholera because he continued to live in the midst of the *coppery* dust and emanations, while others lost their lives by fleeing from the, as they thought, infected atmosphere of the workshop."

The views and recommendations of Dr. Burq, as to the uses of copper in cholera epidemics, were largely indorsed by those who gave them any thoughtful attention, and copper chains and plates were worn by many

people. But proofs accumulate from allopathic
sources. The "miasmatic, animated beings," mentioned by Hahnemann, recognized and defined by
Koch as microbes peculiar to Asiatic cholera, are con-
signed to sure and swift destruction, both within and
without the patient, by the use of the *salts of copper!*
During the epidemic of 1884, in France, the fact was
developed that the views of Dr. Burq had gained the
support of such great allopathic lights as Claude Ber-
nard, Littre, Marchal de Calvi, Baron Larrey and Pro-
fessors Charcot, Luys and Dumontpallier.

The latest theory to account for the fact of prevention is, that "salts of copper, taken in doses of from
15 to 20 centigrammes a day, albeit for weeks, harm
neither man nor warm-blooded beast, but slaughter
low organisms." Paul Bert, of Paris, leading all others
in support of what the London *Lancet* facetiously calls
the "Cupric Saviour," asserted that "these salts remain for some time in the system, and particularly in
the liver, and are innocuous, unless taken in large
doses, and the person whose liver is saturated with
copper may defy cholera."

Without waiting to determine whether it be to the
electric, magnetic, mechanical, antiparasitic or homœopathic relationship to the morbific agency in cholera
that the credit of protection is due, the practitioner is
very fully justified in exhibiting *Cuprum* as the best of
all prophylactic agents.

Insoles for the shoes, made of very thin sheet copper, worn with only a thickness of ordinary hose between them and the feet, are more effective and more

convenient than plates or chains elsewhere worn. The triturations of Cuprum, such as made in our pharmacies, should be taken, night and morning. The writer has prescribed the second decimal, in three-grain doses, with the happiest results, in every cholera epidemic in America except the first.

Therapeutics.—If Samuel Hahnemann, in his long lifetime, had done no more than to point out the real character of Asiatic cholera, with the best methods, public and personal, for its prevention, and the most efficient means for the treatment of its subjects, he would have earned a place among the world's greatest benefactors.

Of all the examples of the need of a general therapeutic law, and all the proofs of the value of the one stated in the terms, *similia similibus curantur*, not one has been so convincing and so glorious as that afforded in the homœopathic conflict with the great destroyer from the Ganges.

The approach of the dreaded monster in the countries of Europe was viewed with consternation, because no line of treatment, suggested by clinical experience or the learning of the medical profession, had been found successful against it. It was reserved for Hahnemann, to whom the relationship that must exist between morbid conditions and those instituted by drugs for their removal had been revealed, to name the remedies for cholera, in advance of any personal experience in the treatment of that disease. The little tract, written by him at Köthen, and published at Leipsic in the autumn of 1831, before he had ever seen a case of Asiatic cholera, pointed out the uses of three remedies, *Camphor, Cuprum* and *Veratrum album,* which have saved more lives in jeopardy from that disease than all other drugs put together. In the entire history of medicine there is not an instance of such clear prevision on the part of a writer, and such faultless clinical confirmation of a therapeutic principle, as shown in that little pamphlet and in the literature of Asiatic cholera since 1831.

Hahnemann advised the instant use of *Camphor* on the appearance of cholera symptoms, one drop on a small lump of sugar every five minutes ; also, in severe cases, the rubbing of the palms of the hands and other parts of the surface with it.

He also recommends an enema of two teaspoonfuls of Camphor-spirits in half a pint of warm water, and the evaporation of some of the same on a hot iron in the room—these latter measures especially when the patient was unconscious or unable to swallow.

When the case had progressed to what is now known as the second stage, characterized by cramps as well as severe vomitings or purgings, he advised the use of *Cuprum* in place of Camphor, repeated every hour or half hour.

He also mentioned *Veratrum album* as an indicated remedy in such cases, used alone or in alternation with Cuprum.

For the typhoid condition, at times following a cholera attack, he recommended the employment of *Bryonia* and *Rhus tox.* alternately ; and, in the last or collapse, stage, Camphor was again his favorite remedy.

While Hahnemann advocated careful individualization in medical practice, he was not slow to perceive that, with a specific cause, ever the same, and symptoms showing little variation in different cases, there could be no occasion for a greater number or variety of remedies.

The logical method that had led him to a sound therapeutic principle, he would not discard in the hour of its greatest need, and he generalized drugs as well as symptoms. The most satisfactory success has crowned the efforts of those practitioners who have met the pathognomonic symptoms of cholera with the characteristic symptoms of drugs, wiping out the former, as by magic, under the pointings of *similia*.

Some writers (generally those who have never faced cholera on the battle-field) have questioned the homœopathicity of Camphor, claiming that it could be indicated only in cases designated as *cholera sicca ;* but a proper examination of the pathogenesis of that drug will show that it is homœopathic, in the first stage of all cases—to the shock, the chilliness, the depression and the faintness of the invasion, and the flushing, the dizziness and the epigastric heat of the reaction. In many, perhaps most, cases the invasion may not be noticed, nor yet the reaction that quickly follows, till nausea, vomiting, purging or cramps set in. If it is a good maxim in financial economy that, "when the pence are well taken care of, the pounds will take care of themselves," it may be as good, in the management of cholera, to say that, *when the first stage is well taken care of, those following will take care of themselves.* Most wisely did Hahnemann recommend the use of Camphor, as an immediate preventive, by all persons obliged to attend upon, or come in contact with, cases of cholera. The vital resistance, properly aroused, may prevent invasion, or overcome attack, before the development of the most serious phases of cholera. And Camphor becomes the chief remedy, again, in the last, or collapse stage, of cholera, when the fury of the fight is past, and the forces of life are failing. It will bring back the pulse and restore warmth when stimulants and artificial heat have failed.

The most thorough and successful use of this remedy, anywhere reported, especially in late epidemics, has been in Italy. Dr. Rubini, of Naples, already referred to, has had an experience with it greater, doubtless, than any other physician.

In 1854–55 he treated 703 cases, losing but two ; Dr. Tripi, of the army in Sicily, under the direction of Prince di Satriano, treated 641 cases, in 1854, losing twenty-five ; Dr. Goth, at Geneva, the same year, treated 841 cases, losing seventy-two—all with the saturated solution of Camphor. The more fully to show the confidence reposed in this remedy, a quotation is here made from a paper in the *Monthly Homœopathic Review* (London), for December, 1884, written by the gifted Dr. Tommaso Cigliano, of Naples. He says: From observations made by ourself, and from those we have, with great care, collected from the aforesaid sources, as well as from the reports of patients who have either cured themselves, or been cured by their friends, with *Camphor*, we are able to draw the following conclusions:

1. In the incubation stage of cholera it is, in every case, successful in preventing the development of the disease.

2. In the early stages, that is, when there are already vomiting and purging, it arrests the disease and prevents its further development in 80 per cent. of the cases in which it is used ; and when it fails in its quickly-abortive action it will still, with perseverance and in increased doses, be the means of bringing on the desired reaction, even when the patient falls into the stage of collapse, provided that he be not disturbed by other remedies. In these cases, the reaction will be developed in from two to six hours, or, more rarely, after from twenty-four to forty-eight hours.

3. In the stage of collapse, it succeeds with almost certainty, provided that no other remedies have been previously used, especially such as laudanum and warm bottles.

4. In women, the reaction is often later in making its appearance than in men.

5. In children, who often reject it by the mouth, it succeeds equally well when rubbed in over the stomach, under the armpits, and over the temples, every half-hour ; and then no internal medicine need be used.

6. In the fever of reaction it is still indicated, two or three doses of one or two drops being given in the twenty-four hours. The fever in the majority of cases lasted twelve hours. In no case, where cure followed, were there any typhoid symptoms.

7. On discontinuing the *Camphor* during the stage of reaction, if there be any relapse, it will be cured by

again resorting to the same remedy. But it is more
prudent not to discontinue its use for some days, even
after apparent recovery; though the quantity and frequency of dose may be reduced, one or two doses a
day being sufficient. Only when each administration
is followed by vertigo or other pathogenetic symptoms
should it be given up.

8. In treating a case with *Camphor*, any other remedy, especially Laudanum, disturbs its beneficent action. Warm baths in a majority of cases are injurious,
and sometimes fatal. So, also, is changing the bed or
linen during any delirium which may precede the
sweating, und during the sweating itself. All these
rules are important in securing good results from treatment.

Dr. Cigliano, after preparing statistics of treatment,
claims that, whereas the mortality under the use of
Camphor, Cuprum, Veratrum album, Arsenicum, and
Carbo veg., and other remedies, after the usual man-
ner of the books, at the best is not less than five or six
per cent., under the use of *Rubini's Camphor* alone, it
may not exceed one per cent., the average rate of
mortality under the use of Opium, Calomel, and other
allopathic drugs being, at the same time over 50 per
cent.

Rubini's Camphor has been given usually in five-
drop doses on small lumps of sugar; in grave cases
it has been given in ten- or fifteen-drop doses repeated
every fifteen minutes.

During the prevalence of cholera in a community,
or in places not far away, prompt attention should be

given to disorders of stomach and bowels, especially to any fulness in head, dizziness, nausea, or diarrhœa. Beside the uses of Cuprum and Camphor, as already indicated, it is necessary, in case of diarrhœa, to employ *Arsenicum*, if the stools are quite liquid, of brownish or greenish color, attended with great thirst, nausea, or a sense of epigastric oppression ; *Croton tig.* or *Ricinus*, when the stools are more liquid, of lighter color, and more copious and gushing.

When the pathognomonic symptoms of cholera are present, and Camphor has failed to prevent the progress of the disease, the following remedies may be successfully employed : *Cuprum met.*, when there are cramps in the limbs, or muscles of the trunk, with or without vomiting or purging, a dose every half-hour ; *Veratrum album*, when there is extreme nausea, active vomiting, gushing, almost involuntary stools, with other grave symptoms. In full and urgent cases, these two remedies have been very successfully used in alternation fifteen minutes apart. When there is great perspiration, with coldness and blueness of skin, and other appearances of collapse, *Camphor* should be given, and if the cramps persist, it may be given in connection with the *Cupram*, say two doses of the former to one of the latter, ten minutes apart.

Other remedies, more or less homœopathically indicated, have been employed with some reported success, such as *Cuprum arsenicosum* (Arsenite of copper), *Phosphoric acid, Secale cor., Hydrocyanic acid, Jatropha,* etc. These, and others which may not be discussed here, are to be considered in exceptional cases, or in

the absence of the more efficient remedies already
named. In cases of excessive nausea, vomiting, and
cramps, the Arsenite of copper promises much.

In regard to *Veratrum*, it should be mentioned that,
after its successful use in homœopathic practice for
more than fifty years in cholera epidemics, an old-
school writer of distinction, Privy Councillor Dr. von
Blödau, of Sondershausen, discovered *Veratrin* to be
an excellent remedy for cholera! " Having frequently
administered strong doses of *Veratrin* as a remedy
against cramps in the calves of the legs during sleep,
and always with success, he concluded that, as it possessed a stimulating influence on the spinal nervous
system, it might restrain the dangers of threatening
symptoms in cholera." Dr. Weber, of Cologne, has
written on the same uses of the drug, but honestly
acknowledged that the Veratrum album had been for
a long time known to homœopathic practitioners as
useful in cholera. In regard to the preparations or
attenuations in which the remedy should be used, the
writer would say that each prescriber should be fully
persuaded in his own mind. His own experience,
which has been ample, taking in every epidemic that
has occurred in America since that of 1832–33, has
given him greater confidence in the lower attenuations,
ranging from the first to the sixth decimal, except in
the case of *Camphor*, when he would use *Rubini's saturated solution.*

In conclusion, on the medicinal treatment of cholera,
it should be said that the most successful practitioner
will ever be the one who, quickly recognizing the foe,

selects and adheres to well-tried remedies, nothing doubting as to good results. He who takes time to search a repertory with a long list of drugs, each of which may have displayed some symptom similar to one characteristic of cholera, will often find his patient fatally collapsed before his individualizing pursuit is satisfactorily ended.

Accessory Measures.—As already intimated, the successful management of cases of cholera does not depend entirely on the drugs administered. The patient should be enjoined to obey certain requirements, among which these are the chief:

1. To lie down at once, and keep still, not rising from the pillow to take the medicine, nor even in the act of vomiting or purging. Let pans and cloths be used so as to receive and remove ejecta promptly and effectually. The matters ejected should be mixed or saturated with a strong solution of sulphate of copper or carbolic acid, and then burned as speedily as possible—in no case being deposited in a cesspool or sewer.

2. To have courage and confidence in what is being done, and a determination to survive the attack. To this end the physician and all attendants should be self-possessed and cheerful, not allowing a word or look of discouragement.

A few sharp words to patient or nurse, or some rallying jests from the physician, have sometimes broken the spell of fear and snatched patients from the jaws of death. The room of the cholera subject is no place for long-faced or undecided doctors and nurses.

3. To take no food till the progress of the disease

is checked; and to abstain from drinks also, taking moderately of pounded ice, when the thirst is intolerable.

4. To keep well under covers, not too heavy, so as to preserve vital heat. Artificial heat, in bricks, irons, bottles of hot water, etc., is of service only in the stage of collapse, after the evacuations and excessive sweating have ceased.

When there is no more vomiting, and a desire for nourishment is expressed, well boiled rice-water or barley-water, seasoned slightly with salt, may be allowed; and when that is well borne and something stronger is desired, beef-tea, without grease may be taken.

Sequelæ.—Patients treated homœopathically seldom have the inflammatory affections, or fever, observed to follow cholera attacks, where opiates, astringents, and more heroic drugs, have been freely used. Should there be any such, they must have the remedies indicated and usually employed for the same conditions arising from other causes.

www.ingramcontent.com/pod-product-compliance
Lightning Source LLC
Chambersburg PA
CBHW022030190326
41519CB00010B/1646